Translucent Gears

RICHARD SILBERG

North Atlantic Books
Richmond, California

Translucent Gears

Copyright 1982 by Richard Silberg

Poems included in this book have appeared in the following magazines: *Phantasm, City Miner,* and *Ball,* as well as in the anthology *Tunnel Road #1.* Several of these poems were originally published in a chapbook, *The Reckoning,* by Oyez Press of Berkeley in 1979.

Typeset in Optima, Souvenir and Eras by Joe Safdie

Published by North Atlantic Books
635 Amador Street
Richmond, California 94805

ISBN 0-938190-06-7 paperback
ISBN 0-939190-05-9 signed cloth

Translucent Gears is sponsored by the Society for the Study of Native Arts and Sciences, a nonprofit educational corporation whose goals are to develop an ecological and crosscultural perspective linking various scientific, social, and artistic fields; to nurture a holistic view of arts, sciences, humanities, and healing; and to publish and distribute literature on the relationship of mind, body, and nature.

I

A SONG FOR THE PIANO PLAYER

Sleazing in a basement
    in Northern California
I hear the black kids
hoot from Harlem
horn fisted graceful
as scary sugar

My Daddy comes
to me moping
a pack rat
torn sheet and rubber mat
in my teen-age thirties
a writer living in bare places

Daddy be my juju
be my African black sceptre
give me courage Daddy
that I be strong
when I face the lion

Down in the black
    ocean streets
I fought psychopathic
cross-wise Sanford
and when his fist
exploded in my head
it was my mother's rage
that rose in me
    and won

Was that the lion
the dragon of my passage?
    or is he here
in this damp winter basement
spiders weaving on the walls
    breathing here
in this gray scuzzy flow
of places
and faces without roots
or flowers
the dragon of failure
of middle age and death?

I hated you then, Daddy
I was ashamed
when I needed someone
to show me how to be a man

We played games instead
        you and I
together like two little boys
travelling on the lit up nerves
        of Manhattan

to Hopalong Cassidy movies
        on Saturdays in Times Square
to the Museum of Natural History
        The Bronx Zoo

You told me stories
shaving in the mornings
about Oscar Schmulewitz
        and his Daddy
their wondrous adventures
travelling by helicopter to Africa
        being swallowed by whales

My slight funny Daddy
cared for by women

A juju for the hollowness
        the concrete and the strangers
a chieftain's amulet and scars
to brave this shapeless river
        of our lives
where we struggle
like little fishes

It's all different now
        an undreamt land
of strange golden hills
swelling like breasts
mountains to the north
ocean to the west

No collapsing black star
        of New York Island
fabulous auras

of show business greats
you booked in the Catskills
        when they were little
of ex-champs
blinking in the Broadway lights
who called you 'Commodore'
and taught me
how to hold my dukes

    no strutting shadows
only bland healthy faces
the wild popcorn boys
I fought in the schoolyard
are junkies or night porters
        squashed candy wrappers
that flutter on other streets

A charm Daddy
a magician's light
to unclench my fists
so the sap can flow
to the middle of my life
so I can stand
in wide horizontal spaces
and listen to the earth

        before my birth
you were a young man
piano player on the road
    I hear you playing
BOOM-CHICK BOOM-CHICK
funny Al
    slick
in lonely towns
Dallas, Rio de Janeiro
making up songs
        for the waitresses

Who you are
a basic party harlequin
only your sensitivity your nervousness
slight clever build
        quick
a woman's man

runner
and a lover
not a fighter

The nervous center
that you couldn't share with anyone
not with Mother or with me
sealed off
in punchlines and babytalk
golf TV laughter
        hard timeless nut of you
dangling inside your seventy years
moustachio moon face and jowls
round belly skinny legs

        gleaming soul nut
unravelling to drop
so soon

I'll store it for my winter
the juju of my father
        sunlight before birth
            the channel
juju of the nervous funny
            talking man
Who I am
conjuring here
in the basement

A nervous funny
talking man
with courage
from my mother

a golden oval
        juju

to let my Daddy go

to feel the cats
the rhythm flow
the black girls
dancing in a row
Marilyn and Sonya
molasses and honey

And when the whale comes
to swallow me
in the ocean Daddy
to know who I am

THE FURRY INTELLECT

for Charles Bukowski

Sometimes a man
stands up
and he's a giant
big enough to crunch
through all the cream
        and beebop

He's got thick hairy
arms
so long they reach
his asshole
which is pink as
        a seashell
and smells like
shit

he sniffs
his finger

he is pleased

Can such a man
be sensitive
the way a bear is
sensitive

      snuffle up
sharp tangy
grubs
      fuck-fuck
into a juicy
she bear
    be
the nasty spark
inside his deepset
eye

      and
still perfect
white crystals
of philosophy?

BLACK BART

I got so down
               giggin'
a hundred fifteen pounds
I weigh a hundred fifty five now
you couldn't tell me from two drumsticks

There's rhythm
   dah dah dah
stays even right through the breaks

There's progression
   CDF CDF
an' like that — progression

An' dynamics — loud an' soft
   ooooo WAAH oooooooo WAAAAH

My dream is to have a recording studio
               in the mountains
music, man, is creation   green

with all these cabins
where the musicians could play

SEPARATE FACES

You were my friend, Harold
sad and rounded
        as a cello
sweet and jumpy
as a thirty six year old
        little boy

You had been violent
and bitten attendants
when you thought you were King
    when you grew 'disturbed'
and fell through
the fractured frozen edges
            in your eyes

That was it
that black spiky shape
that you fell in on
            each time
vibrated upon
        like a bug
pinned in ice contemplation

I wanted to give you my eyes
    so you could see
that you're excitable
gay as a magpie
that there are orbits
around your black ice
where you could spin
        and bloom
just like me

But you grew aggressive
        coming at me
        sucking out
with your hands
and your yellow teeth

That spiky black shape
stood in the cracked lens
            of your mind

Gleaming and bulging
            slyly
more
    you wanted
                more
who had had nothing
trying to grab through
the red lining of skin
and bone between our two faces

THE AUDITION

This cat jumps up to my table
    an actor he says
    out of our sly past
running around each other
in and out of pastel doors

a black cat
serpentine
lithe obsidian blade

       mumble mumble
      auditions
    mumble seven hundred bucks
    from his old lady
    Hollywood mumble
    talent and beauty

the round conviction of his face
circles galaxies of want
resplendent thrones
caged white leopards

I want you to come out
and listen to this speech
      Othello

    in the park
    he readies himself
    becomes
    a planted broomstick
children play behind his head
    while he declaims
    down a long murky hall

sunlight searching in the grass
needles into
blue-veined tiny flowers
with dark blue
smoky threads of pollen

He has beauty too
honey light in his darkness
curved as Nefertiti
on the prow of her ship of Afterlife
curved frozen black beauty

    waiting in his sunlit hallway
for the turbaned moor
who will never appear

      money
    stardom

for the pig of light to come rooting at his soul

OUTSIDE IN THE TREES

His sideways head crouches
cunning as an animal
among the brilliant facts
the sun on passing cars
trees rustling by the empty house
Why? Why? he says
I wish I knew
Why did I do it?
this shadowy graceful crouching
moving splitways
between mirrors
sucking
in its dark tight nut
      between

Why?
when I loved her?
he asks again
as if God stood in the sunny street
as if combination locks
clicked
somewhere outside in the trees

THE RECKONING

Back in my small bare room
                at 3AM
my thunder hammer
hung up on the wall
unknown Chinese
middle age
sat glinting
jovial
intimate
as an orangutang

he wasn't phased
when I bellowed
and beat my chest
but leaned to me
solicitously,
"So you have the blues?",
                he said

Stretched as I was
against the red city sky
my monkey face
spun prayer wheels
in his eyes

JADE PACES

Chinese philosophers give us
the red muscles
on which the sky rolls
    dragons
that coil through the swaying trees

their black silk sleeves conjure
terrible significance
each thing speaking
gorgeously garbled
a tower rising
in gold light
towards high flat Heaven

These are dreams
in love's dull spaces
embers
glowing just behind the spine

TRANSLUCENCE

Strange crackling in the kitchen
and there on my table
among the dazed breakfast dishes
     sat a bird
a lost crumb of the magic world

     curved yellow beak
     iridescent brown body
rocked back on his heels
like a drunk on a lamp post
breathing triple-time
shallow and fast

then he flapped up
jaggedly, without sense, into the window
    — CRACK! —
I felt the pain of it
beak in my mind
wings I had forgotten from childhood
he was lost in a castle of mind
the pale green kitchen walls
the strange clear hardness
that rose up between him and the sky

Catch him, catch him
little jewel brother
before he shatters himself
fluttering at the glass in terror
like a rag in a wind

   when I slipped him out the window
beating softly in my hand
he flew

the clean streak of his flight
opened my flesh
flying away
translucent as fruit flesh

SLANTING PARTY IN THE AFTERNOON

The shadows of little girls
mingle on the wall

They're sitting on the toolshed roof
secret as creamy cats
above their mommies
whose skirts ride up bare thighs
unfurling
muscle honey lines
brown fat
prancing and stamping
to the music

Men and women dance
a bulging Breughel mass
a sunlight volvox breathing
joyous bubbles
of red sunlight

Slenderly
musician brothers
boogie in a basket
their beards and straw hats bending every way
        dreaming up white snakes
                of music

rising up
with birds and steeples
from the Earth
so vast and delicately green
sprinkled with cavorting centaurs
rolling
half in darkness half in light

UNIVERSAL CITY

She unwound me so cleanly
she was the wind
        an awkward soft dustdevil
that whirled across an open place

More a wizard than a woman
        pink gums
        pocked face
blue silver and blonde

And left me with myself
the men and women
that I knew
hungry couples
single hunters

tentacles
of an anemone
contracting on its blackness

A shining in her eyes
concerned and soft
    unwavering

'You're good,' I said
'a bright blue stone'
'And you're the picador,' said she,
'the beam of light that makes me flash'

Then we were stars

all our murmuring heads

a scattering of stars
that twinkled
in the empty universal city
        which is night
        which is a flower

LONG-LEGGED BIRD

Today
I woke up
startled from sleep

      and caught

my slim soul
a long-legged bird
wading out
into a foggy marsh

A Chinese landscape
      to be sure
but with real
microscopic vision
scum light
of water bugs
and pollywogs
spiralling
in green pools
    at the shores

living
fog water

I was dartlooking
for danger
red morsels
to feed goggle eyes

a publisher's check
a woman to love

A LETTER FROM FLAHERTY

Driving to work this morning   it was bright
and clear   the day seems liquid or crystal
   sometimes

     driving down route 2   doing about sixty

on the road in front of the van   suddenly
  were the shadows

     they seemed to be straining to be born

  for the instant

   black black birds
   in the asphalt sky

 and passed
    beneath the wheels

  before thinking
 or feeling could begin

REFLECTING SURFACE

Biking up
the cat is lying in the road
asleep my nerves pray
with the familiar dropping chill
until the face turns
to look at me

A cartoon
red popping eyes and tongue
a funny death
laughing quietly in my bones
sweet sleeping
little pussycat
death

And then gone
biking down the road
towards my morning coffee
singing rock and roll

it can't be touched
we know it only
by the exit marks
where something has passed
the squashed cat face
is depthless
a reflecting surface only
like light from the road ahead
rising up to strike our eyes
at a peculiar angle

STAR FUR

Connie slung Sam Dillon on her back
and stumped off
    laughing
      to the river

she swayed away
from our drunken light
so sweetly prim and sliver

   our high
bloodcoursing party
the trees rising up
around us
like shouts
   slow explosions

Sam Dillon rode her chalice body
brown muscle sweet river
that night
swinging
like a compass needle
on the world's silken flanks

and reminded me
of my sunlit young mother
the myth of my mother
her animal body
   like earth
the star fur of sky gods
who made love to the earth
I remembered the volcano
we bushmen crawled out of
rapt little monkies
crawling into time

THE KINGDOM

My momma's face
arched pursing sea
of sunlight
before my memory

Driving through the midst of Long Island
in such violent summer money hammering
we could practically see the new houses
        growing
visions in prefabricated dreams

Driving out to see the ancestral land
the plot that Granma bought
for $1300 in 1929

It was the middle of my life
my chasm of need
to turn this family treasure
        into money

Love talk
private love talk
as we drove
the family
who they had become
orthodontists, CPA's
our generations
expanding out of NYC
pizza parlors in LA
condominiums in Florida

        And me
the brightest promise
golden boychik
an unknown dreamy bum
small enough to slip
        without racket
through the wasp waist
of the relentless hour glass

        Fresh cooling breeze of Oyster Bay
and we were there
pacing on our property

It was a narrow joke of land
40' by 100'
at the sandy shoulder
of a bulldozed road

Vertiginous
all crystal tears
my ancestors spread like ink
disappearing into dark maps
centuries
and shtetls
        this worthless strip
all that was left
of Granma
seabed
powdered pink
            her high swooping voice

Crystal march
on my own mysterious land
        narrow grass
crossroads from nowhere
going nowhere

A transient moment
in electric history
our pyramid of internal life
tottered on its nose

    How my life
had been a dream
digging inside
while interest accrued
and lines zigzagged on charts

Squandered, squandered
my momma's golden eyes
these tatter-assed humors
of the lost
        immemorial Kingdom

GRANMA

it's an illusion
that I connect you with trains
engines of death and departure
coming as I did
at the end of your life

Bucharest
your Mongol cheekbones
painted angels
whirling on the hatbox

the words branch out
like spines
        like catwalks
over breathing oxen
of machines

there's the body of a woman there
shining like headlights
she died in the subway
she was a baby
cooked at the center of the earth

AT THIRTEEN
after a poem by W.S.

I tunneled with slim Tony
clever and mystic
darkly melting as a candle
down beneath the high school
digging for treasure
    through an air vent
towards a distant twinkling light
our Indies
    the girls' locker room
where Hope Finney
would stand revealed
impossibly naked
swaying
like a palm tree at the mind's end
    red coconuts
monkeys crawling wickedly
    in her bush

We labored
   panting through the maze
and it was the furnace room instead
hairy Gale shovelling
    truly
like a lumpen ape

Running out
cackling hysterically
Tony always in the lead
   doubled
scrambling in the tunnel
    to be born
I hit an overhanging pipe
   and saw stars

MOVING

Lion heads
with rings through their mouths
the black brass cannon
in the Indian Museum

terror holes
playgrounds
the bridge and the river

Without Larry
things led nowhere
they just sat still
ended in brick or stone

Larry was foxy
knew the ways in
skinny
the color of corn
his belt was too long
and he could run
in his baggy pants
his legs swinging wide
as the clapper of a bell

he could hang by his heels
he could disappear
into tenement walls

He was the explorer
I was the philosopher

sitting on the rocks
down by the Hudson
I told him things I read in books
and his eyes popped round
                with wonder

how the Earth had spun
from the fiery sun
how waves of life
had filled the seas
and crawled on land

giant roaches
dinosaurs
and ape men

around us
in the shadows
the rats
slunk big as cats
the condoms floated
        with the empty bottles

We talked of growing up
and how we'd move away
I would go to College
he would join the Navy
while we watched the freighters
steaming down the river
a half mile out
delicate
their magic toy rigging
moving against the New Jersey shore

WALKING

Not since childhood
this dream
frozen sea
pink waves
like mashed potatoes and catsup
           on my plate
like the blood froth tissues
of a farflung body
I am alone
walking the stiffness of these waves
keyed up
pulled
        half in terror

there on the horizon
leaps the whale
frozen
curving and trumpeting
sly on his head
his cloudy buzzing eye

MY FATHER'S SHOW

I remember hiding there in the dark
          in the Scotch
          in being a kid
while they danced
out front in the spotlights
a hulking dark man
in a black tuxedo
whirling his small partner overhead
    curves of her torso
molded in pink sequins

They were just a warmup act
for the singer and the comedian
I had met them backstage
aging show business athletes
         never going to be stars

but my feeling of terror
the impossible perfection
         of being a man
of cigarette smoke   neon lights on the cars
of key rings   sex   and bank accounts
coalesced on them
the dark bull man
        hunched and stiff
holding up the arching salmon body
of the woman to the light

TUNNELING

He steps on the train
a blind boy
with gummed sunken eyes
"Here—to the right," says his guide,
and he turns the wrong way
"Oh—to the right!," he exclaims
    turning back gaily
high forced good sport voice
that I hear inside me
    the boy that I was
posing bravely
Not like this boy—not at all
this young beardless man really
pale as fungus
hair matted
brows jutting
    over the vestigial sinking of his eyes
But I heard him
inside
And here
down the stairs
a sandy girl
turning
confused
great blue eyes
    of startled flight
I feel her
    hesitating
the wind blowing
awkwardly lifting her woolen skirt
around her thin legs
lost
in a world of forms
tangled chalky flowers
separate
in the sunlight
all darkly rooted
    millionfold

IN THE WINGS

Tip-toeing for love on his lumpish calves
forging the shields helmets and swords
Noel's fulminating white shape
                        crouches
to peep through the crack in the dressing room door
His pointed black eyes spark
creases leap on his brow
        little angel dog face
        sprightly shining anguish

THE ANGLES

Then she said the spinal thing
the tattoo of a dagger on his belly
        her father
a tough small man
bright bird's eyes
                joking to the end
and when they operated for his cancer
they cut into the tattoo
That sealed it for her
red symbol
the fixation

and her mother
a Grant Wood pitchfork character
with plump rouged cheeks
                before the church
Dead dull deadly
Modesto petty
while she a blonde wind feathered
shy eyed goat girl
but a tough narrow hipped little pixie
her father's hoarse humor
breathing
behind the angles of her face

GOD'S KITCHEN

You've been digging in the body for treasure
      the rich golden black
         corrupting body
      riding the dragon spine
         down and away

Then you flip off to Writer's Land
         flying imagination
         out of your prison
            up and out
to the aerial view

Oaf!
you go down grinning each time
thinking you've got the keys
and can run off
      jingling like a thief

But it's life life
LIFE!
flowing by you
light streaming through your head
         which is cracked
like a plate of golden egg yolks
dropped in God's Kitchen

ROCKS IN MY GARDEN

It's terrible
    Daddy
    Philip Whalen
your happiness is terrible to me
condemning me to hope in life
to the possible
      golden
      shooting

Self pity is so sweet
      moony life in death
to be a jackal
gnawing at poet bones

But you!
in your fat Buddhahood
rubbing your shaven head
like a prayer bead
kneading your head
like a kernel of corn

strenuous life
prying at rocks in the garden

BABA

"Will you draw me a picture of a man?"
                                    he asked
          white-garbed
an Indian
turning his abrupt dark eyes to me
where we sat at separate tables
          on the boulevard
Then he explicated my drawing
to a colleague from the West
"The exaggerated musculature
stiffness in the back
          like a fighting dog
and yet he has no face
he has no fists."
Dark center of his eyes
eyes of many faces
sounding like bells
"Your fight is foolish.
You will die." they said
many many
murmuring green cycles
hot stars burning in cold space
not merciful
          and not unkind

THE PAINFUL MYSTERY OF LOVE, ETC.

Ridiculous to dither about
blood circuses
thickets bursting with flowers
heart threads
glistening in star patterns
like mitosis

There was just your tough hot smile
sad worried spic eyes
my glowing eyes of many colors
amazing skinny body, etc.

And in the clinch
we climbed together
on the same golden rod
slowly
gently
making honey

climbing in the afternoons

before we came down

DIFFERENT KINDS OF WORK

If I were an active man
farmer   householder
I would weed this place
clear
beautify
lay in a garden
and make it bloom
But
another kind of worker
I take our backyard inside
the way a child sucks a candy
                    voluptuously
I note the coltishness
of the overturned metal tub
      caress
or menace?
of the spruce
groping at the bayberry bush
how the rocks hum calmly
            in black earth
Nothing is changed
when my work is done
and God
        patiently
    goes on hiding

MEAT

Mark playing jazz piano
in private room at school
      soundproof dots

like Renaissance anatomist
        Vesalius
in midnight cone of light
         (Rembrandt
browning with age)
dissecting this body
       I say

white midnight
the pale architecture of body
     the fingers exploring
   at the center

CHERRIES

instant little child woman
Arletha
Minnie Pearl
round in your forties
as a hill of black cherries

that passion you gave me
in stilted cliches
"You fuckah!
  You fuckah!
    Oooohh! . . . I can feel the juices."
you lived in them truly
like birds
in bird-locked trills

like your living-room
of old lady lace
lamps and the tables
the lacquered beringletted face
              of Jesus
on the kitchen calendar
rolling his dark sugar orbs
to Heaven
over bold flat script
              that said
"Joe's Laundry".

JANE DEAD AT 32

like light on shaken water
she walked
              slowly
wobbling

her tiny frame
puffed
        fat belly
pollywog in metamorphosis

My sickness is my own
this process going on inside me
my shtick
my trip
but sometimes she wondered
Do I want to be sick?
am I responsible
for my sickness?

small woman
sweet
seeming vulnerable
            but inside
something was flat
as a hammerhead
pounding at life

lovers
pained writing
good jobs
doctors
psychotherapists
est

           at 14
she did a tough bopper's
stiff-legged
shiney-eyed dance

hard and young

           she floats back
dancing
in the black marsh light
           of changing things

LOOKING INSIDE

i

to have the jackal at heart
emptiness
whirling
striking in desire
the sun face of teeth and burning
moon face of loneliness
dark
      bound apart

ii

looking inside or outside
i the point of lens
a clear nothingness
      humming
        where the beams converge

iii

a cat prowls deeply through the grass
radio sounds in the trees
sunlight
      birds flickering
the house is made
of dark
      shingled wood and masonry
     voices
daisies stemming from a glass decanter

iv

i am waiting for the flame
for these round stones
fingered smooth and slick
to burst in flame
      inwardness
how it
laps
     upon its stony bowels
        stone spine
        stone eyes
            and wings

SPLIT SECOND

Maybe it was the intersection
   space between halves
Leon talking to his daughter on the phone
"There's something I want to tell you—
                 I love you"
and Georgia O'Keefe's eyes
       pierced
         suddenly opened
but it was the sky that poured out first
how huge
   ripping off veils of white twilight
             a wide suction of space
and the tree caught me
   moving by surprise
What had I forgotten
         laughter?
            terror blooming?
moving many dark arms
   with the movements of the wind

TALL BLOSSOM

I was
bellowing
in the shadows
of a vast cathedral

when you came stoking along
                    on cool stilts
like an old cut
from the Coasters

Your long silver head
                possessed itself
churning it out
in starry slow motion

hit a penny
hit a penny
        bump a nickle twice
bump it jellybean nice

eyes of a newborn baby
dark steel blue
receding to
chalk cliffs and the sea
a little horse in California
                    prison time

I love you
    I love you   I said
tall silver blossom
ultra-Rilkean beebop angel

No time
no time   you said
I'm a bruised blues singer
        light and gay
I've acted in England
and danced ballet

No time
no time
See how the leopard
        eats my flesh
I'm a little shorn lamb
birds sing
        in my hair

Then I knew who you were
alone
flowering
between the sea and the sky
brilliant heads
        of our lovelessness

my painful sister
dancing
in the vast cathedral

DESERT

You've got to
                    keep coming
on the other hand
there is the black glove
casual
          Botticelli placed
over the cunt

reclining in a train robbery
in cactus country

on the other hand
the poem of your funkiness
and wrinkles
your shopworn beautiful speed trip

fast as an antelope, McClellan!

men and women slam out
180° bad
in Berkeley-Oakland
              it's like
people are murdering each other at this incredible
RATE

    life

                  clangs on
                  TILT

    the machine hits
with its incandescent tits

there is the desert
and cactus country
        Love
white as a bone
        with all the delicate hues
              and overtones

SONG OF THE MIRROR CHILD

Time in our minds
is a round blue lake
outside the atoms tick
our skulls will echo when we die
but in my mind, Momma
my diving blue mind

        Your voice lilts
as you read to me
        floating
    a happy balloon
over the butter and syrup
        of my pancakes
where I dwell
in the red orange forest
                of my childhood
you are the sun in the sky
One day I'll be a great man
I'll take you away
                and educate you
so my refined lovely Momma
will be free
            to shine

one day
            one day

Santa's potent cherry nose
was glowing in the aircourt
How does Santa always know?
airplanes and women
and every time I spin
            to catch
            the elfin forces
in a twinkling
            always gone

projectiles hurtling in blue tubes

We are sitting
in this alien junior four apartment
in cannibal Queens

Mom and Pop and I
and they are old
and I am middle-aged
the jet planes whistle overhead
dividing time like swords of light

God
everything is shrunken
to such tiny terms
empty lakes of Mother
            Pop and me
empty moon crater lakes
and money
bitterness
of how it stands in our eyes
whistling towards death

All so
            incidental
this present
like the twinkling
            of the spokes
of a huge wheel

some strange
mistake
your face so sad and helpless
ringed in fat

Momma
you're like my puffy little girl
            now
you've lived for me
and I have lived for
            Truth?
                  Magic?

Childhood miracles.
Don't worry, Momma
I'll snatch you off the tracks
blue
            vanishing

I'll be your hero, Momma
melt you down
dance you out
      orange and black
where the tigers turn to butter
biting their tails
around Black Sambo's tree

TWO CENTS

What is this poetry?
lightning through Frankenstein flesh
common crumbling coin
the red clay cliffs
run down the lemurs   run down howl
pouring myself into emptiness

What was this Merwin poetry? lemurs
                    in shifting tones
the side of the cube waving off
the daylight roaring

my own mean-spirited wish
to murder green flames

What do you open to?   Small grey place
                Some common market
                    place

LSD CLASSMATES HOME TO ROOST

Precisely as the nodes of a vibrating string
eerie muscles of a fate
across the white river
                of the 60's
         we find ourselves
                here
            staring through new faces

JUST FRIENDS

How can it be born between us?

I'm full of wrinkling
             nets of light

envious
       would seek to kill her

       for her tall red pants
and James
blond motorcycle boys
                  shorn
                  punk look

Love
if not flesh
why not bones?
why not stone
arches?

her eyes
are deep wells
where the spirit wrinkles
in unheard shapes

       like animals
playing in bloodfused light

I'm full
as a golden cup
my heart
hurts

TO FLASH

Meat carriages in moving landscape
dark brown pointed sky
                of your eyes
rolling sea of cunt earth
                        hips

moving meat carriages
                anxious radii
lives moving
pain
ecstasy
        moving

Goodbye, mama
goodbye little god face

pure apples
        peaches
    on the steps

sun trees
flashing by jagged steps of light

there where our bodies burn
        dark nodding saints

        burning to terrible flash
        just to flash there

before   after

JUICE

Word wheel
off the hairy sun
brilliant slow motion bursting
of the word wheel

        the leaves

spry tongues

the pouring down cat people
curving on their spines

softly
       softly

trembling juice

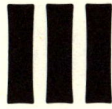

BELMONT '78

I

New York
New York again
driving in gritty yellow air
through the clear time passing
Danny and I
who had known each other half our lives
the furtive princely sons of Jewish mothers
still running for our pleasure

Driving through the Barrio
116th St. and Spanish Harlem
on our way to the Belmont Stakes

as if we drove down the bones of old romances
"I used to do social work in this neighborhood, Rich . . .
Used to make love to my clients . . ."
Danny hanging sidewise, snake eyes, furioso
his cigarette clenched between his teeth
romance of blackness
fierce rolling women
soul, electric salsa, broken glass
straight-on surging of that poor angry blood
so seductive to us
          interior middle-class darters
only we weren't kids anymore
but men scrabbling in deep water
and outside the romance was gone
of lives caged to a few square blocks
flattened like scrap or baling wire
in the infernal machines
empty store fronts
stale wine breath
strutting and hard laughter
at the core of a dying star

Danny and I
talking women, money, races
in our closed moving cell
like two heads of the same fear
he hanging tight as a wart
in the city of his birth
I escaped a continent away
veined by phone calls and jet planes
to the anxious stone beebop of the city

And here
crossing the bridge
Randall's Island floating in the East River
Manhattan State Hospital passing on our right
        squat fearsome buildings
rising tan and blunt as hog's teeth

That had been another of our romances
we Jewish schoolboys working as psychiatric attendants
in the brilliant twisted inner jungle
the zoo
garbage hospital
for the poor people of Manhattan Island

sadness and terror of who was still there
which of our patients
as if the time had stopped
fifteen years
of jellied animal eyes, devil eyes, dead eyes
imploding
flesh falling to the still helpless center

and behind it the whale, itself,
stretching to the north and south
theatres, embassies, publishing houses
the tiny jagged buildings like erector sets
unnaturally clear in the sulfurous air
countless tiny people juxtaposed
like planets stacked in cubes
        Leviathan
the humped island of Mind

fading to our backs
as we chattered over the racing form
fading
as we chattered on the past
and I remembered dreams I had as a child
of the humming beast
seen from the inside
all the heads connected
to the subway dragon spine
power junctions rumbling
          down beneath the island
the lemur furnaces
          the secret dynamos

II

Belmont Park spread
          flying for miles
in a peaceful ellipse
wide green circus
of fields and ponds
          candy-striped tents
groves of trees and jets roaring over
thousands of well-heeled people
twittering
rustling their racing forms
betting on the horses

As if the day shifted on translucent gears
of beauty and quick money
and the animals
how they floated onto the track behind the bugle
nervous flaring thoroughbreds
stepping up and back before the stands
anchored to their firm squatty little exercise ponies
floated out in colors round the curve
jittering one by one into the starting gates
and "They're off!"
the hard clear straining
tiny monkey men in silks
     like a Grecian frieze
     an Assyrian lion hunt on TV

Danny and I
still getting by
love boys once again
here at the hot center
glowing at each other
as if our lives leapt
upon these pearls of pleasure
        strung through emptiness

losing and collecting under the sky
in radioactive green Flushing, NY
JFK airport whistling over
beer and food
convivial New York tippling from coast to coast

All building toward the Stakes
effervescence
expectation
of Belmont '78
the third and longest test
of Alydar and Affirmed
Stevie Cauthen the wonder boy
in his shot to sweep

And then
        unbelievably
just before the race
"The Concorde! The Concorde!"
screeching
scrowling up into the sky
a fierce white bird
as big as the moon
so big and piercing
I felt I could hold it
the rippling of its radar muscles
giant alphabets clicking together
squeeze some shape from History
streaming by
in every direction
        like light

III

When they burst from the gate
there came a roar
it was like the rolling of a wave
and they were running free
Affirmed in front
Alydar a few lengths behind
running easily through the first turn

The day grew suddenly deep
the horses galloping in that wide green field
sound surging
as thousands of people urged them
praying money
        veins and pop eyes
Danny was screaming beside me
and I felt myself lifted out
into a huge crackling bell of energy

Alydar starting his move along the back stretch
gaining on Affirmed
        gaining
he pulled even coming into the clubhouse turn
and our minds went white

Alive inside the whale
the cruel beast of History
this race created by New York
as surely as Rome made the circuses
winners and the losers
the rich floating
upon crushed heads of the poor
to build these instruments of glory

Alydar and Affirmed running together
neck in neck
sculpted heads locked
in a pure Alexandrine straining
galloping the last half mile
in a pitiless stone stretch
like the pumping of some giant heart

And Affirmed won
again
by a nose

Straight up
Danny and I
hugging each other
punching each other
          electrified
          out of our lives

And when he rode back into the winner's circle
this frail piqued eighteen year old in pink silks
                    with his arms full of flowers
he was magnified on TV
and we screamed with the crowd
in that beating hollow
fused together shrieking in release
          for Stevie Cauthen
winner of the Triple Crown